4th International Conference on Epigenetics and Bioengineering (EpiBio 2020)

Online
9-11 October 2020

ISBN: 978-1-7138-2776-4

Printed from e-media with permission by:

Curran Associates, Inc.
57 Morehouse Lane
Red Hook, NY 12571

Some format issues inherent in the e-media version may also appear in this print version.

Printed with permission by Curran Associates, Inc. (2021)

For permission requests, please contact AIChE
at the address below.

AIChE
120 Wall Street, FL 23
New York, NY 10005-4020

Phone: (800) 242-4363
Fax: (203) 775-5177

www.aiche.org

Additional copies of this publication are available from:

Curran Associates, Inc.
57 Morehouse Lane
Red Hook, NY 12571 USA
Phone: 845-758-0400
Fax: 845-758-2633
Email: curran@proceedings.com
Web: www.proceedings.com

TABLE OF CONTENTS

Dynamic Imaging of Nascent RNA Reveals General Principles of Transcription Dynamics

Conference: International Conference on Epigenetics and Bioengineering

Year: 2020

Proceeding: 4th International Conference on Epigenetics and Bioengineering (EpiBio 2020)

Group: General Submissions

Session: Emerging Technologies

Authors:

Larson, D., National Cancer Institute

The stochastic activities of RNA polymerase and the spliceosome are responsible for the heterogeneity of the abundance and isoform composition of mRNA in human cells. However, the dynamics of these megadalton enzymatic complexes working in concert on endogenous genes have not been described. Here, we establish a quasi-genome-scale platform for observing synthesis and processing kinetics of single nascent RNA molecules in real time. We find that all observed genes show transcriptional bursting. We observe short active periods and variable inactive periods ranging from minutes to days. The heterogeneity in inactive times gives rise to the widely observed "noise" in human gene expression and explains the distribution of protein levels in human tissue. We derive a mathematical model of regulation that relates transcription, chromosome structure, and the cell's ability to sense changes in upstream signals and predicts that hypervariability is largely dynamic and does not reflect a stable biological state.

Genome Replication As Origin of Regulatory Variation and Epigenetic Drift

Conference: International Conference on Epigenetics and Bioengineering

Year: 2020

Proceeding: 4th International Conference on Epigenetics and Bioengineering (EpiBio 2020)

Group: General Submissions

Session: Emerging Technologies

Authors:

Morival, J., University of California, Irvine
Ren, H., University of California, Irvine
Akhtar, N., University of California, Irvine
Laszik, N., University of California, Irvine
Nie, Q., University of California, Irvine
Read, E., University of California, Irvine
Downing, T., University of California, Irvine

Molecular heterogeneity is emerging as a critical feature of multicellular life. While single-cell analyses have revealed the existence of cell-to-cell variation in the levels and activities of the molecules responsible for gene regulation, the source of such variation is still poorly understood. Cytosine methylation is a highly conserved epigenetic modification that plays an important role in mammalian development and its occurrence within phenotypically uniform cell populations is often variable even at the same genomic location. We recently developed a new sequencing method (Repli-BS) that enables analysis of methylation heterogeneity across cytosine residues within newly replicated strands of DNA over time. Using this method, we discovered that much of the methylation heterogeneity observed within HUES64 human embryonic stem cells (hESCs) is temporal in nature and associated with DNA replication (Charlton et al., *Nat. Struc. Mol. Bio.* 2018). More recently, we used our hESC Repli-BS dataset to establish kinetic rate parameters that numerically reflect the â€œspeedâ€☐ at which individual cytosines achieve steady-state methylation levels after being replicated (Busto-Moner et al., *Plos Comp. Bio.* 2020). Here, we employ bioinformatic analyses to explore how properties of post-replication DNA methylation dynamics relate to well-established features of the genome and the broader chromatin landscape. Our preliminary findings reveal that unique patterns of methylome replication associate with distal regulatory regions throughout the genome, enrich for cytosine residues dynamically methylated between stem cells and germ layers, and coincide with the location of stem cell-specific transcription factor binding and chromatin architectures. We also find correlations between sub-cell cycle kinetics in DNA methylation and the divergence of bulk methylation patterns observed during multiple cell generations and natural aging. Taken together, our studies suggest that (epi)genome replication may act as an important source of (temporal) regulatory variation in hESCs while, simultaneously, conferring susceptibility to epigenetic drift throughout the human lifespan.

Genome-Wide Mapping of Protein-DNA Interaction

Conference: International Conference on Epigenetics and Bioengineering

Year: 2020

Proceeding: 4th International Conference on Epigenetics and Bioengineering (EpiBio 2020)

Group: General Submissions

Session: Emerging Technologies

Authors:

Henikoff, S., Fred Hutchinson Cancer Research Center

We recently introduced new tools for high-resolution genome-wide chromatin profiling and have applied them to study chromatin organization and dynamics in a variety of model systems, including yeast, flies and mammals. Among the questions addressed are: How do transcription factors find their binding sites in DNA packaged into nucleosomes? How are nucleosomes depleted from gene regulatory regions? What is the relationship between nucleosome deposition pathways and chromatin deregulation in disease? Our findings illustrate ways in which chromatin dynamics can play a central role in regulating gene expression and silencing.

Measurement of DNA Methylation and Nuclear Organization with Nanopore Sequencing

Conference: International Conference on Epigenetics and Bioengineering

Year: 2020

Proceeding: 4th International Conference on Epigenetics and Bioengineering (EpiBio 2020)

Group: General Submissions

Session: Emerging Technologies

Authors:

Razaghi, R., Johns Hopkins University
Timp, W., Johns Hopkins University
Steensel, B. V., Netherlands Cancer Institute
Peric, D., Netherlands Cancer Institute
Lee, I., Johns Hopkins University

Nanopore sequencing has enormous potential in epigenetic applications; unlike traditional sequencing-by-synthesis technologies, it can distinguish covalently modified nucleotides directly through their modulation of the electrolytic current. We can take advantage of the long read lengths generated by nanopore sequencing to precisely call methylation patterns, and to obtain phased methylation information across the genome. The promise of nanopore sequencing has already been demonstrated by the accurate calling of 5-methylcytosine in a CG context using hidden Markov models (nanopolish) and neural networks (Taiyaki/Megalodon). We have used trained models on other modified (non-canonical) nucleotides, focussing here on dam-mediated adenosine methylation (N6-mA).

Methylated adenosine sequencing has applications in bacterial epigenetics as well as exogenous labeling in mammalian systems. In particular, the damID method uses a dam enzyme fused to a protein of interest, e.g. nuclear lamina, resulting in labeling DNA proximal to the protein. We are able to detect this N6-mA modification simultaneously with 5mC, providing two distinct epigenetic signatures along ~10kb single DNA molecules. To train our model distinguishing methylated from unmethylated adenosine, we sequenced both completely unmethylated, fully dam-methylated (N6-mA), and dam (N6-mA) and M.SssI (5-mC) methylated E. coli and human genomic DNA, where dam methylates adenine residues in a 5â€™-GATC-3â€™ context and M.SssI methylates cytosine residues in a 5â€™-CG-3â€™ context. We applied this model to mammalian cells treated by damID to characterize lamin-associated domains (LADs). LADs identified using microarray probes showed on average significantly higher methylation profiles than other regions in our pipeline. Applying this model we can simultaneously measure phased nuclear positioning information (LADs) with DNA cytosine methylation (5-mC) using long reads.

Controlling Cancer By Epigenetics: Are We Ready for the Prime Time?

Conference: International Conference on Epigenetics and Bioengineering

Year: 2020

Proceeding: 4th International Conference on Epigenetics and Bioengineering (EpiBio 2020)

Group: General Submissions

Session: Epigenetics in Human Health & Disease

Authors:

Verma, M., National Cancer Institute

Epigenomics provides essential tools not only for therapeutic intervention but also as predictors of drug response. Unlike alterations in genetic components where the structure of a gene is changed (such as mutations, deletions, additions, single nucleotide polymorphism), epigenetics causes changes in gene expression without altering the genetic sequence. Furthermore, several epigenetic changes are reversible, and because of this, drugs have been developed that can reverse the epigenetic changes associated with different cancers. These epigenetic drugs target DNA methyltransferases, histone deacetylases and proteins associated with post-transcriptionally modified histones. Epigenetic regulators represent potential therapeutic targets as they often have binding domains that lend themselves well to small molecule inhibition. Resistance to cancer therapy is an ongoing challenge, and there is a need to seek out non-traditional approaches such as targeting epigenetic components for cancer therapy. Factors that contribute to the development of resistance to targeted therapy and how epigenetic regulators can be utilized to overcome this resistance will be discussed. Examples of several cancer types will be presented for which conventional targeted therapy has limited effects, but improved treatment outcomes are observed following epigenetic reprogramming of gene expression patterns, including modification of histones and DNA. Results from clinical trials have indicated the effectiveness of epigenetic drugs as cancer therapy, mainly when administered in combination with traditional anticancer drugs. These promising data suggest that epigenetic drugs have great potential in controlling and treating cancer.

Decoding the Epigenomic Regulation of Neuronal Plasticity

Conference: International Conference on Epigenetics and Bioengineering

Year: 2020

Proceeding: 4th International Conference on Epigenetics and Bioengineering (EpiBio 2020)

Group: General Submissions

Session: Epigenetics in Human Health & Disease

Authors:

West, A., Duke University

Transcriptional enhancers are regulatory DNA elements that underlie the specificity and dynamic patterns of gene expression. Over the past decade, large-scale functional genomics projects have driven transformative progress in our identification of enhancers and our understanding of the mechanisms by which they regulate gene expression. However, much remains to be learned about how enhancers orchestrate the biology of specific cellular processes including neuronal differentiation and plasticity. Neurons are dynamic cells that respond and adapt to stimuli throughout their long postmitotic lives at least in part by changing their programs of gene transcription. We have been investigating how chromatin regulation at enhancers coordinates these transcriptional adaptations in order to illuminate the roles of enhancers in neuronal biology. I will discuss the evidence and proposed mechanisms for the regulation of chromatin state and structure at enhancers in postmitotic neurons. I will also share data from our ongoing studies, in which we are using CRISPR/dCas-mediated epigenome editing to directly test the functional consequences of enhancer chromatin regulation for both neuronal gene expression and the adaptation of neuronal function.

Neuronal Activity-Induced BRG1 Phosphorylation Regulates Enhancer Activation

Conference: International Conference on Epigenetics and Bioengineering

Year: 2020

Proceeding: 4th International Conference on Epigenetics and Bioengineering (EpiBio 2020)

Group: General Submissions

Session: Epigenetics in Human Health & Disease

Authors:

Wu, J., University of Texas Southwestern Medical Center

Neuronal activity-induced enhancers drive the gene induction in response to stimulation. activity-regulated gene expression plays a central role in both short-term neural responses and long-term neural adaptation. Altered expression causes changes in neuronal function and connection that may lead to behavior defects and brain disorders. Mammalian SWI/SNF-like ATP-dependent chromatin remodeling BAF complexes, which contain core ATPase subunits BRG1, modulate chromatin structures and regulate transcription. BRG1 is a key node in the autism spectrum disorder gene network. Mutations in BAF subunits lead to neural developmental disorders. Previously, we and others have shown that neuronal BAF complexes are important for activity-regulated gene expression and synaptic development and plasticity. Here, we demonstrate that BRG1 regulates neuronal activity-induced enhancers. Upon stimulation, BRG1 is recruited to enhancers in an H3K27Ac-dependent manner. BRG1 regulates enhancer basal activities and inducibility by affecting cohesin binding, enhancer-promoter looping, local H3K27Ac levels, RNA polymerase II recruitment, and enhancer RNA expression. Furthermore, using a proteomic approach, we identified a serine phosphorylation site in BRG1 that is induced by neuronal activities and is sensitive to CaMKII inhibition. BRG1 phosphorylation affects its interaction with several transcription co-factors including NuRD co-repressor complexes and cohesin, possibly modulating BRG1 mediated transcription outcomes. Using mice with knock-in mutations, we showed that non-phosphorylatable BRG1 fails to efficiently induce activity-dependent genes, whereas phosphomimic BRG1 increases the enhancer activities and inducibility. These BRG1 phospho-mutant mice displayed anxiety-like phenotypes and altered responses to stress. Therefore, our data reveal a mechanism connecting neuronal signaling to enhancer activities through BRG1 phosphorylation. Our study also provides significant insights into the function of chromatin remodeling complexes in neural plasticity and brain disorders.

3D Epigenome Remodelling in Endocrine Resistant Breast Cancer

Conference: International Conference on Epigenetics and Bioengineering

Year: 2020

Proceeding: 4th International Conference on Epigenetics and Bioengineering (EpiBio 2020)

Group: General Submissions

Session: Nuclear Organization

Authors:

Achinger-Kawecka, J., Garvan Institute of Medical Research
Valdes Mora, F., Garvan Institute of Medical Research
Loi Luu, P., Garvan Institute of Medical Research
Stirzaker, C., Garvan Institute of Medical Research
Clark, S., Garvan Institute of Medical Research

Endocrine therapy resistance frequently develops in estrogen receptor positive (ER+) breast cancer, but the underlying molecular mechanisms are largely unknown (1). Here, we use high-resolution chromosome conformation capture (Hi-C) to show that 3-dimensional (3D) chromatin interactions both within and between topologically associating domains (TADs) frequently change in ER+ endocrine resistant breast cancer cells (2). We provide an important connection between genetic and epigenetics changes in endocrine resistant breast cancer by integrating chromatin contacts with whole-genome sequencing data, which highlights the role of resistance-associated genetic variants at CTCF-bound anchors in promoting differential interactions. Ectopic chromatin interactions are also preferentially enriched at active enhancers and promoters and ER binding sites and are associated with altered expression of ER-regulated genes, consistent with dynamic remodelling of ER pathways accompanying the development of endocrine resistance. Importantly, loss of 3D chromatin interactions often occurs coincidently with DNA hyper-methylation and loss of ER binding. Alterations in active (A-type) and inactive (B-type) chromosomal compartments are also associated with decreased ER binding and atypical interactions and gene expression. Together, our results demonstrate for the first time that 3D epigenome remodelling is a key mechanism underlying endocrine resistance in ER+ breast cancer providing new avenues for therapeutic targeting.

1. Stone, A. *et al.* DNA methylation of oestrogen-regulated enhancers defines endocrine sensitivity in breast cancer. *Nature communications* 6, 7758 (2015).

2. Achinger-Kawecka *et al.* Epigenetic reprogramming at estrogen-receptor binding sites alters 3D chromatin landscape in endocrine-resistant breast cancer. *Nature Commun* 11, 320 (2020)

A Cohesin Cancer Mutation Reveals a Role for the Hinge Domain in Genome Organization and Gene Expression

Conference: International Conference on Epigenetics and Bioengineering

Year: 2020

Proceeding: 4th International Conference on Epigenetics and Bioengineering (EpiBio 2020)

Group: General Submissions

Session: Nuclear Organization

Authors:

Carico, Z., University of North Carolina
Dowen, J., University of North Carolina Chapel Hill

The cohesin complex functions in the spatial organization of the genome into DNA loop structures. While cohesin is known to occupy enhancers, promoters and CTCF sites, the molecular mechanisms through which cohesin is loaded onto chromatin, localizes to these specific genomic sites, then forms and maintains loops are not fully understood. In yeast, targeted mutagenesis and suppressor screens identified the hinge domains of Smc1 and Smc3 as being required for cohesin loading onto chromatin and the topological entrapment of sister chromatids during DNA replication. It is not known whether the cohesin hinge also regulates these functions in mammalian cells or participates in metazoan-specific functions in interphase chromatin organization. Investigation of cancer sequencing databases revealed a hotspot mutation in the SMC1A hinge at amino acid residue arginine (R) 586 in leukemia. We used CRISPR/Cas9 genome editing to generate murine embryonic stem cell (ESC) lines with an endogenous tryptophan (W) mutation at amino acid 586, as seen in cancers. Transcriptomic analysis of SMC1A-R586W ESCs reveals global dysregulation of the transcriptional programs, with reduced expression of pluripotency-controlling TFs, activation of differentiation programs, and misexpression of oncogenes. Strikingly, RAD21 ChIP-seq indicates that SMC1A-R586W reduces cohesin enrichment at promoters and enhancers, but not at CTCF sites. Hi-C analysis reveals altered genome organization, including a reduction in short-range, intra-domain DNA contacts. This work indicates that the cohesin hinge participates in the cohesin loading reaction at promoters and enhancers, and/or promotes the localization of cohesin to these elements. These findings further suggest that cohesin mutations can contribute to oncogenesis through selective loss of cohesin at transcriptional regulatory elements.

Epigenetic Control of Cell State By Chromatin Looping

Conference: International Conference on Epigenetics and Bioengineering

Year: 2020

Proceeding: 4th International Conference on Epigenetics and Bioengineering (EpiBio 2020)

Group: General Submissions

Session: Nuclear Organization

Authors:

Dowen, J., University of North Carolina Chapel Hill

Mammalian gene expression is regulated by various epigenetic mechanisms including DNA methylation, the histone code and the spatial organization of the genome. Knowledge of how patterns of gene expression are maintained or dynamically change during development is key to understanding control of cellular identity. Embryonic stem cells are a fascinating model system for studies of cellular identity since they can both self-renew and give rise to all cell types of an organism. The looping of chromatin by cohesin and CTCF is an important regulator of embryonic stem cell gene expression programs. Altered function of cohesin, and its regulators, disrupts proper gene expression programs and leads to differentiation. Changes in cohesin dynamics and localization are associated with aberrant organization of the genome within the nucleus. Cohesin-dependent chromatin loops serve as insulators of transcriptional activity and thereby link genome structure and genome function.

Biochemical and Cellular Libraries Reveal Cancer-Associated Histone Mutations That Perturb Nucleosome Structure and Inhibit Cell Differentiation

Conference: International Conference on Epigenetics and Bioengineering

Year: 2020

Proceeding: 4th International Conference on Epigenetics and Bioengineering (EpiBio 2020)

Group: Poster Session

Session: Poster Session

Authors:

Bagert, J., California Institute of Technology
Mitchener, M. M., Prin
Lemiesz, A. E., The Rockefeller University
Dul, B. E., Princeton University
Wojcik, F., Princeton University
Nacev, B. A., The Rockefeller University
Feng, L., The Rockefeller University
Allis, C. D., The Rockefeller University
Muir, T. W., Princeton University

Recent whole genome sequencing data mining efforts have revealed thousands of mutations in histones, the DNA-packaging proteins in eukaryotic genomes, across a wide range of cancer types. These occur in all four core histones in both the unstructured tail and folded globular domains and remain largely uncharacterized. Here we used two high-throughput approaches, a DNA-barcoded mononucleosome library and a humanized yeast library, to profile the biochemical and cellular effects of these mutations. We identified cancer-associated mutations in the histone globular domains that enhance fundamental chromatin remodeling processes, nucleosome sliding and histone exchange, and are lethal in yeast. In mammalian cells these mutations promote cancer phenotypes, upregulating cancer-associated gene pathways and inhibiting cellular differentiation by altering expression of lineage-specific transcription factors. This work represents the first comprehensive functional analysis of the histone mutational landscape in human cancers, and leads to a model in which histone mutations that perturb nucleosome remodeling can contribute to the development of cancer phenotypes.

Dose-Dependent Activation of Gene Expression Is Achieved Using CRISPR and Small Molecules That Recruit Endogenous Chromatin Machinery

Conference: International Conference on Epigenetics and Bioengineering

Year: 2020

Proceeding: 4th International Conference on Epigenetics and Bioengineering (EpiBio 2020)

Group: Poster Session

Session: Poster Session

Authors:

Lu, D., UNC Eshelman School of Pharmacy
Hathaway, N., The UNC Eshelman School of Pharmacy
Chiarella, A., University of North Carolina, Chapel Hill
Butler, K., Icahn School of Medicine at Mount Sinai
Jin, J., Icahn School of Medicine at Mount Sinai
Gryder, B., NIH
Khan, J., NIH

Gene expression can be activated or suppressed using CRISPR/Cas9 systems. However, tools that enable dose dependent activation of gene expression without the use of exogenous transcription regulatory proteins are lacking. We created a technology termed chemical epigenetic modifiers (CEMs) which can regulate the expression of target genes by recruiting components of the endogenous chromatin regulatory machinery, eliminating the need for exogenous transcriptional effectors. The system has two parts: catalytically inactive Cas9 (dCas9) in complex with FK506 binding protein (FKBP), and a CEM consisting of FK506 linked to a molecule that interacts with cellular epigenetic machinery. We show that CEMs upregulate gene expression at target endogenous loci up to 20-fold or more depending on the gene. We also demonstrate dose-dependent control of transcriptional activation, function across multiple diverse genes, reversibility of CEM activity, and genome wide specificity of our best-in-class system.

Dissecting the Role of DNA Methylation in Tuberculosis-Induced Immunological Exhaustion Using CRISPR/Cas9-Based Epigenome Editing

Conference: International Conference on Epigenetics and Bioengineering

Year: 2020

Proceeding: 4th International Conference on Epigenetics and Bioengineering (EpiBio 2020)

Group: Poster Session

Session: Poster Session

Authors:

Guerra-Resendez, R. S., Rice University
Nishiguchi, T., Baylor College of Medicine
DiNardo, A., Baylor College of Medicine
Hilton, I., Rice University

Tuberculosis (TB) infection results in 10 million cases and 1.5 million deaths a year, making TB the worldâ€™s leading infectious cause of mortality, and one of the top ten causes of mortality overall. We have recently shown that TB patients display generalized immunological exhaustion and dampened T Cell-mediated immune responses, and that these deficits are correlated with DNA hypermethylation at immune signaling pathway genes (e.g. *IL12RB2* and *IFN-Î³*) in TB patient T Cells relative to healthy household controls. Despite these correlations, the molecular mechanisms by which DNA hypermethylation can illicit inhibited host immunity remain unclear.

To clarify the role of focal DNA hypermethylation in TB-induced immunological exhaustion, we applied epigenome editing using dCas9-DNMT3A/3L to target DNA hypermethylation to specific loci in the *IL12RB2* and *IFN-Î³* genes based on differentially methylated regions (DMR) identified in TB patient-derived immune cells. Our preliminary studies in HEK293T cells reveal that *IL12RB2*and *IFN-Î³* gene expression decreases in response to targeted DNA hypermethylation and suggest that forced DNA methylation at these loci may be directly causal for TB-mediated immune exhaustion, even after eradication of *M. tuberculosis* from patients. Additionally, to understand how global levels of DNA methylation alter immune cell phenotype, we used CRISPRa (dCas9-VPR) and CRISPRi (dCas9-KRAB) to up- and downregulate, respectively, the *DNMT3A* and *TET1* genes.

Ongoing work is aimed at investigating the immune phenotype disruption using flow cytometry-based multi-dimensional immune profiling and mycobacterial growth inhibition assays. We expect that the precision control of DNA methylation in immune cells will illuminate how specific changes in DNA methylation perturb immune function and open new avenues for a generalized therapeutic intervention for immune exhaustion, which is a hallmark of other infections and several human cancers.

Mechanical Regulation of Histone H3 Modifications Via Matrix Rigidity during TGF?1-Induced Epithelial-Mesenchymal Transition

Conference: International Conference on Epigenetics and Bioengineering

Year: 2020

Proceeding: 4th International Conference on Epigenetics and Bioengineering (EpiBio 2020)

Group: Poster Session

Session: Poster Session

Authors:

Sankhe, C. S., The Pennsylvania State University
Gomez, E. W., The Pennsylvania State University

Epithelial-Mesenchymal Transition (EMT) is a biological process that is important in development and has been implicated in carcinogenesis and fibrosis. During EMT, cells lose apical-basal polarity, experience gene expression changes, and gain enhanced motility. The extracellular matrix (ECM) surrounding cells plays a major role in regulating cellular functions and studies have shown that increased ECM rigidity promotes gene expression changes associated with EMT. In addition, evidence supports epigenetic remodeling during EMT, however, there is limited knowledge available on how the mechanical properties of the ECM control chromatin and nuclear architecture along with histone modifications in the context of EMT. Here, we synthesized hydrogels with stiffnesses spanning that of normal and diseased mammary tissue and examined the effect of matrix rigidity on mammary epithelial cells during transforming growth factor (TGF) Î²1-induced EMT. Using experimental assays such as immunofluorescence staining and western blotting, we find that matrix rigidity modulates bulk levels of histone H3 modifications during TGFÎ²1-induced EMT. These modifications are associated with transcriptional activation and silencing. Furthermore, cell contractility inhibitors modulate the levels of histone markers in response to increasing matrix rigidity and TGFÎ²1 treatment. These findings suggest that matrix rigidity contributes to epigenetic remodeling and subsequent gene transcription in the context of the EMT process. This, in turn, will help to identify approaches to control EMT processes in pathological conditions such as fibrosis and cancer metastasis.

Analysis of Archived Human Heart Specimens Uncovers Cardiomyopathy-Specific Genomic Signatures

Conference: International Conference on Epigenetics and Bioengineering

Year: 2020

Proceeding: 4th International Conference on Epigenetics and Bioengineering (EpiBio 2020)

Group: Poster Session

Session: Poster Session

Authors:

Bhattacharyya, S., UT Southwestern Medical Center Dallas

Direct evaluation of primary human tissue is unparalleled for translational investigation of organ-specific physiology and disease. Aside from readily accessible (e.g. skin) and commonly biopsied (e.g. tumor) tissues, most translational evaluation remains challenging. Regulatory restrictions associated with organ procurement and extensive tissue fixation of autopsy specimens present additional hurdles. Consequently, government-sponsored organ procurement efforts (e.g. National Disease Research Exchange and GTEx) have improved access to difficult-to-obtain human tissues, but still contain relatively few specimens. Alternatively, RNA and DNA amplification is typically required to obtain transcriptomic and epigenomic information from archived specimens. Here we describe a highly optimized, versatile, and straightforward procedure to simultaneously purify RNA and DNA from fresh, fresh frozen, or fixed human heart tissue. To demonstrate the utility of our method, we generated high-quality RNA-seq and ATAC-seq datasets from primary heart tissue derived from rejected donor hearts and autopsy specimens. Next, we obtained high-resolution chromatin accessibility maps from three separate forms of human cardiomyopathy: ischemic, dilated, and hypertrophic. Detailed analysis of epigenomic data uncovered cardiomyopathy-specific genomic signatures. Gene Ontology and motif enrichment analyses identified relevant biological terms and transcription factors to suggest mechanistic differences of each cardiomyopathy subtype. Furthermore, the chromatin accessibility maps also guided retrospective categorization of difficult-to-diagnose cardiomyopathy subtype in additional patients. Additional chromatin accessibility maps unbiasedly classified pre-left ventricular assist devices (LVAD) and remodeled (post-LVAD implantation) hearts from same individual (N=2). Cardiovascular Genome Wide Association Studies (GWAS) variants landing on subtype specific chromatin signatures correlated to disease specific impact on gene regulation. As an effort towards furture diagnostic application, we established a simple ATAC qPCR methodology to detect disease subtypes from limited tissue. Taken together, our new methodology utilizing stable epigenomic features provide a robust pipeline for translational discovery and drive disease etiology using archived specimens over conventional RNA-seq or microarray-based disease prediction approaches.

Chromatin-Mediated Alternative Splicing Regulates Cocaine Reward Behavior

Conference: International Conference on Epigenetics and Bioengineering

Year: 2020

Proceeding: 4th International Conference on Epigenetics and Bioengineering (EpiBio 2020)

Group: Poster Session

Session: Poster Session

Authors:

Lombroso, S.
Heller, E. A., University of Pennsylvania
Xu, S. J., University of Pennsylvania

Drugs of abuse are known to cause remodeling of neuronal cell populations that persists long after the drug is no longer present. It is well established that cocaine epigenetically regulates gene expression, yet mechanism(s) by which cocaine-induced epigenetic modifications regulate *alternative splicing* is largely unexplored.

We tested the hypothesis that H3K36me3 directly mediates alternative splicing in brain in the context of preclinical models of drug abuse. We first quantified concomitant splice-site enrichment of H3K36me3 and differential isoform expression in mouse brain tissue following cocaine intravenous self-administration (IVSA). We showed that cocaine IVSA led to enrichment of H3K36me3 at splice junctions of differentially spliced genes and that alternative splicing serves as a key transcriptional mechanism in three brain reward regions (NAC, PFC, and VTA). To validate that enrichment of H3K36me3 was causally linked to alternative splicing, we then overexpressed SET2 in the NAc. We processed infected tissue using single sample sequencing (S3EQ) to generate H3K36me3 ChIP- and RNA-seq data in order to identify H3K36me3-associated alternative splicing events. We found that SET2-driven enrichment of H3k36me3 was sufficient to increase alternative splicing in a subset of neuronal genes that were also mediated by cocaine self-administration.

Next, we tested if H3k36me3 enrichment was sufficient to alter cocaine behavior and found that SET2-mediated H3K36me3 enrichment in NAc increased cocaine seeking in both Conditioned Place Preference (CPP) and IVSA. In addition, we showed these effects can be rescued with a SET2 inhibitor. From these gene lists, we identified the splice factor, Srsf11, as highly enriched amongst both cocaine and SET2 spliced genes. Using CRISPR dcas9-SET2, we applied targeted epigenetic editing to establish that H3K36me3 functions directly in alternative splicing of Srsf11. Epigenetic editing of H3K36me3 at Srsf11 was sufficient to drive alternative splicing of Srsf11 and enhance cocaine CPP and IVSA.

Genome-Wide Transcription Burst Size Quantification through 4sUDRB Sequencing

Conference: International Conference on Epigenetics and Bioengineering

Year: 2020

Proceeding: 4th International Conference on Epigenetics and Bioengineering (EpiBio 2020)

Group: Poster Session

Session: Poster Session

Authors:

Beckman, W., Universiteit van Amsterdam
Verschure, P., Swammerdam Institute for Life Sciences
Westerhoff, H., University of Amsterdam

Transcription bursting is a substantial source of gene-expression noise for many genes, contributing to phenotypic heterogeneity and potentially driving both physiological and pathological processes such as tumorigenesis and drug resistance. Characterising bursting dynamics at a genome-wide scale has been incomplete however as it has been restricted to inferring bursts in mRNA formation computationally from the heterogeneity of mRNA levels in single cell transcriptomic data. Here, we measured the bursting of transcription for *GREB1* in single MCF-7 cells and used this to calibrate genome-wide measurements from 4sUDRB sequencing. Comparing across the genome, we find a significant trend in which the transcription burst sizes range from 1 to ~200 and attributes to ~85% of the steady state expression level, while the burst frequency attributes minimally. Individual genes deviate strongly from this trend and engage both in anomalous burst size and in anomalous burst frequency. We find that presence of the TATA box or Inr sequence within gene promoters significantly predicts a larger burst size, as does a promoter associated YY1 transcription factor binding motif. We also reveal several epigenetic marks to be strongly associated with transcription burst size, such as H3K79me2 and H3K14ac. This method paves the way to accurately quantify transcription dynamics genome-wide in various cell types to elucidate the mechanistic role of cell-type specific epigenetic gene regulation.

Human Cerebral Organoids Reveal Early Spatiotemporal Dynamics and Pharmacological Responses of UBE3A

Conference: International Conference on Epigenetics and Bioengineering

Year: 2020

Proceeding: 4th International Conference on Epigenetics and Bioengineering (EpiBio 2020)

Group: Poster Session

Session: Poster Session

Authors:

Sen, D., North Carolina State University
Keung, A. J., North Carolina State University
Drobna, Z., North Carolina State University
Voulgaropoulos, A., North Carolina State University

Angelman Syndrome is a complex neurodevelopmental disorder characterized by delayed development, intellectual disability, speech impairment, and ataxia. It results from the loss of UBE3A protein, an E3 ubiquitin ligase, in neurons of the brain. *UBE3A* is paternally imprinted in neurons and despite the dynamic spatiotemporal expression of UBE3A observed in rodents and the potential clinical importance of when and where it is expressed, its expression pattern in humans remains unknown. This reflects a common challenge of studying human neurodevelopment: prenatal periods are hard to access experimentally. In this work, human cerebral organoids reveal a change from weak to strong UBE3A in neuronal nuclei within 3 weeks of culture. Angelman Syndrome hiPSC derived organoids also exhibit early imprinting of paternal UBE3A, with topoisomerase inhibitors partially rescuing UBE3A levels and calcium transient phenotypes. This work establishes human cerebral organoids as an important model for studying UBE3A and motivates their broader use in understanding complex spatiotemporal and epigenetic molecular events and neurodevelopmental disorders.

Influence of Hydrogel Stiffness and Laminin 511 on Human Neural Stem Cell Potency

Conference: International Conference on Epigenetics and Bioengineering

Year: 2020

Proceeding: 4th International Conference on Epigenetics and Bioengineering (EpiBio 2020)

Group: Poster Session

Session: Poster Session

Authors:

Philip, D., The University of Akron
Willits, R., The University of Akron

One of the major constituents of the ECM in the nervous system is laminin (LN); and so far 5 alpha, 3 beta, and 3 gamma units of the LN chain are known, resulting in various combinations of the LN isoforms. However, studying the effect of LN isoforms in vivo is challenging, due to overlapping effects of LN chains. To simplify these effects, LN isoforms can be investigated in vitro on a synthetic hydrogel mimicking neural tissue stiffness and functionalized with the isoform of interest. Currently, LN isoforms, such as LN 511 and others, are used as 2D culture coatings to maintain pluripotency of mouse and human embryonic stem cells, and induced pluripotent stem cells (iPSCs). LN 511 is also used for deriving mature and functional neurons from these iPSCs. As LN 511 has been shown to sustain mature and functional neurons from hPSCs, this gives rise to the question would LN 511 also maintain proliferation and expansion of NSCs? The NSC niche is naturally present in the subventricular zone of the adult brain and is made up of fractal structures, which are mainly composed of LN 511 isoform. Therefore, we hypothesize that LN 511 is a key factor in sustaining the defining characteristics of NSCs to proliferate and self-renew on substrates of optimal stiffnesses. Appropriate microenvironmental cues such as functionality and stiffness must be incorporated to better mimic the NSC niche in vitro. In this study, one of the most abundant LN isoforms present in the NSC niche, LN 511, will be incorporated onto synthetic substrates. The LN 511 will be tethered onto hydrogels with a stiffness ranging from soft, as preferred by neurons, to stiff, as preferred by glia. Thus, we aim to investigate how substrate stiffness, tethered with LN 511, influences NSC behavior through mechanotransduction signaling pathways.

Influence of Touch-Spun Fibers Diameter on Neurite Outgrowth and Schwann Cell Migration

Conference: International Conference on Epigenetics and Bioengineering

Year: 2020

Proceeding: 4th International Conference on Epigenetics and Bioengineering (EpiBio 2020)

Group: Poster Session

Session: Poster Session

Authors:

Cavanaugh, M., Northeastern University
Willits, R., The University of Akron
Becker, M., Duke University
Motta, C., The University of Akron
Asheghali, D., Duke University
Silantyeva, E., The University of Akron

Introduction: Peripheral nerve repair across large gaps represents a clinical challenge. One solution is the development of implantable polymeric scaffolds as alternatives to autografts. Nanofibrous scaffolds are ideal for tissue engineering applications because their dimensions are similar to the components in the extracellular matrix and they can be seeded with cells to facilitate regeneration. We will investigate the effects of nanofiber diameter on a Dorsal root ganglia (DRG) whole tissue system and on a Schwann cell (SC) single cell system.

Materials and Methods: *Fiber preparation:* Aligned touch-spun PCL fibers of four different fiber diameter groups ranging from 300 nm to 900 nm were fabricated. The touch-spun samples were characterized using a field emission scanning electron microscope (SEM). *Cell Culture:* DRG were obtained from Sprague Dawley rats and cultured on surfaces for 7 days. The effect of fiber diameter on SC emigration and neurite length was examined using immunohistochemistry. Lastly, primary SCs were used to examine single cell migration behaviors on the nanofiber and control groups including cell alignment, velocity (µm/min) and diffusivity. Statistical differences between groups were determined using ANOVA, with $p<0.05$. Data are presented as the mean ± standard deviation.

Results and Discussion: Analysis of neurite length showed 700 nm fibers had significantly longer neurites than the plastic control. SC emigration was significantly more elongated along the all fiber groups compared to the controls. Single SC migration showed that cell alignment increased with fiber diameter. Also, fiber samples showed diffusive results regardless of diameter. Lastly, 900 nm fibers had significantly higher Instantaneous Velocity (µm/min) compared to the other fiber groups. Thus, fibers enhanced alignment of SC migration but did not produce a directional effect. Overall, larger fibers promote regeneration but, more work is needed for further examination of the mechanisms involved.

Kennedy Pathway Regulation By Mir-16-1/15a Cluster Family during T Cell Activation

Conference: International Conference on Epigenetics and Bioengineering

Year: 2020

Proceeding: 4th International Conference on Epigenetics and Bioengineering (EpiBio 2020)

Group: Poster Session

Session: Poster Session

Authors:

Urena, F., Univeristy of Hawaii
Hoffmann, P., University of Hawaii At Manoa
MA, C., Univeristy of Hawaii at Manoa
Horgen, D. F., Hawaii Pacific University
Hoffmann, F., Univeristy of Hawaii at Manoa

The Kennedy pathway in the endoplasmic reticulum is vital for the synthesis of phosphatidylethanolamine (PE) that is incorporated into cell membranes and lipid-based signaling molecules. The three steps of the Kennedy pathway are catalyzed by ethanolamine kinase, CTP-phosphoethanolamine Cytidyltransferase, and selenoprotein I (SELENOI). Our data suggest that two of the three enzymes (ethanolamine kinase and SELENOI) included in this pathway are upregulated during T cell activation as a mechanism for increasing PE, to support increased membrane synthesis and lipidated signaling proteins. Moreover, we have analyzed micro-RNAs (miRs) and found that miR-16-1 and miR-15a, which are linked in the genome and are co-expressed, have sequences predicted to target ethanolamine kinase and SELENOI. Thus, we hypothesized that levels of miR-16-1/15a decrease upon T cell activation to allow increased expression of ethanolamine kinase and SELENOI, thereby driving the increased synthesis of PE. Using real-time PCR and western blot analyses, we found that both miR-16-1 and miR-15a decreased, while levels of ethanolamine kinase and SELENOI increased in mouse primary T cells upon T cell receptor (TCR) activation. This coincided with increased levels of different PE species as measured by LC-MS. Overall, these data support the model of miR-16-1/15a regulation of the Kennedy pathway, in which miR-16-1/15a levels are higher in naïve T cells as a means to keep PE synthesis low. Upon TCR engagement, these miRs are reduced and all three enzymes are up-regulated, thereby allowing increased PE synthesis required by activated T cells.

Matrigel Solution Whole Brain Cerebral Organoids for Increased Throughput

Conference: International Conference on Epigenetics and Bioengineering

Year: 2020

Proceeding: 4th International Conference on Epigenetics and Bioengineering (EpiBio 2020)

Group: Poster Session

Session: Poster Session

Authors:

Estridge, R. C., North Carolina State University
Keung, A. J., North Carolina State University
Mischler, A.
Rao, B., North Carolina State University

Human cerebral organoids have become a popular model for studying neurodevelopment and diseases as they can be readily generated from embryonic and induced pluripotent stem cells. Established whole brain cerebral organoids require a process of Matrigel, a gelatinous mixture of extracellular matrix proteins and growth factors secreted by Engelbreth-Holm-Swarm mouse sarcoma cells that promotes polarization and supports organoid buds, embedding. However, this tedious, most time-intensive part of whole brain cerebral organoid generation process restricts the throughput, limiting the number of samples and/or conditions that can be tested. To address this, we instead cultured whole brain cerebral organoids in various concentrations of Matrigel solution and lengths of time. The Lancaster protocol was followed until the Matrigel embedding step. A small amount of Matrigel was then mixed with cerebral organoid differentiation media, added to the organoids, and the organoids were placed on an orbital shaker. The organoids were then imaged over time using a brightfield microscope and their size calculated using ImageJ. The organoids cultured in the Matrigel solution exhibited faster growth and larger size than embedded organoids. Moreover, some organoid conditions exhibited large, fluid-filled cavities. Additionally, organoids at days 25 and 50 were immunostained for maturation (SOX2 and TUJ1), cell type (MAP2, TBR1, FOXA2, FOXG1), and cell death (Caspase-3) markers. The Matrigel solution organoids displayed organized structures and similar cellular diversity to the embedded control. The larger, fluid filled organoids exhibited increased cell death. Using Matrigel solution to culture human whole brain cerebral organoids will allow for expanded research avenues of human neurodevelopment.

Nanobody-Mediated Control of Gene Expression and Epigenetic Memory

Conference: International Conference on Epigenetics and Bioengineering

Year: 2020

Proceeding: 4th International Conference on Epigenetics and Bioengineering (EpiBio 2020)

Group: Poster Session

Session: Poster Session

Authors:

Van, M., Stanford University
Fujimori, T., Stanford University
Bintu, L., Stanford University

Targeting chromatin regulators to specific genomic locations for gene control is emerging as a powerful method in basic research and synthetic biology. However, many chromatin regulators are large, making them difficult to deliver and combine in mammalian cells. Here, we developed a new strategy for gene control using small nanobodies that bind and recruit endogenous chromatin regulators to a gene. We show that an antiGFP nanobody can be used to simultaneously visualize GFP-tagged chromatin regulators and control gene expression, and that nanobodies against HP1 and DNMT1 can silence a reporter gene. Moreover, combining nanobodies together or with other regulators, such as DNMT3A or KRAB, can enhance silencing speed and epigenetic memory. Finally, we use the slow silencing speed and high memory of antiDNMT1 to build a signal duration timer and recorder. These results set the basis for using nanobodies against chromatin regulators for controlling gene expression and epigenetic memory.

Potential of Microsecond Pulse Electric Fields (µsPEF) for Epigenetic Manipulation and Breast Cancer Treatment

Conference: International Conference on Epigenetics and Bioengineering

Year: 2020

Proceeding: 4th International Conference on Epigenetics and Bioengineering (EpiBio 2020)

Group: Poster Session

Session: Poster Session

Authors:

Safaei, Z., Rowan university
Thompson, G., Rowan University

Microsecond pulsed electric fields (ÂµsPEF) have the potential to treat cancer by initiating biochemical cascades. Cell death is induced by electroporation leading to imbalanced intracellular ion concentrations, such as [Ca^{2+}] [1-3]. When each binding site of Ca^{2+}-dependent protein, calmodulin, is saturated with Ca^{2+}, calmodulin undergoes a conformational change that permits it to interact with and activate a diverse set of enzymes, especially the kinase CaMKII [4], which act further downstream. If cell death is not initiated by msPEF exposure, perhaps exposed cells respond via (in)activation of other signaling pathways. For example, CaMKII phosphorylates histone deacetylases such as HDAC4, causing export of HDAC4 from the nucleus â€" thereby disrupting its role in genetic transcription and regulation. Thus, proliferation, growth and development of breast cancer cells can be â€œswitched offâ€☐ [5,6].

In this study, we stimulated nucleocytoplasmic shuttling of HDAC4 in MCF7 cells by applying 300 V for 100 ms with varying numbers of square-wave pulses (1, 10, or 50) at 1 Hz repetition rate. We postulate that increased Ca^{2+} uptake due to msPEF exposure activated calmodulin and CaMKII, leading to HDAC4 translocation outside nuclei. Disruption of proliferation of MCF7 breast cancer cells was manipulated via msPEF exposure-induced inhibition of HDAC4.

References

[1] Mi Y et al., *IEEE Trans Dielectr Electr Insul*, 2009; 16(5):1302â€"1310.

[2] Cemazar, M et al., *Electro- Magnetobiol*, 1998; 17:263-272.

[3] Thompson, GL et al., *J Biomed Opt*, 2014; 19:055005.

[4] Prevarskaya, N et al., *Phil Trans R Soc B*, 2014; 369:20130097.

[5] Backs, J et al., *Mol Cell Biol*, 2008; 28(10):3437-3445.

[6] Li, Y and E Seto, *Cold Spring Harb Perspect Med* 2016:a026831.

Programmable Human Histone Phosphorylation and Gene Activation Using a CRISPR/Cas9-Based Chromatin Kinase

Conference: International Conference on Epigenetics and Bioengineering

Year: 2020

Proceeding: 4th International Conference on Epigenetics and Bioengineering (EpiBio 2020)

Group: Poster Session

Session: Poster Session

Authors:

Li, J., Rice University
Mahata, B., Rice University
Escobar, M., Rice University
Goell, J., Rice University
Wang, K., Rice University
Hilton, I., Rice University

Histone phosphorylation is a ubiquitous post-translational modification that allows eukaryotic cells to rapidly respond to environmental stimuli. Despite correlative evidence linking histone phosphorylation to changes in gene expression, the lack of targeted technologies has hindered the ability to establish the causal role of this key epigenomic modification within native chromatin. To address this technological gap, we built a programmable chromatin kinase, called dCas9-dMSK1, by directly fusing nuclease-null CRISPR/Cas9 to a hyperactive, truncated variant of the human MSK1 histone kinase. Targeting dCas9-dMSK1 to human promoters results in increased target histone phosphorylation and gene activation and demonstrates that hyperphosphorylation of histone H3 serine 28 (H3S28ph) in particular plays a causal role in the transactivation of human promoters. Furthermore, we use dCas9-dMSK1 to show that H3S28ph influences the acetylation status of histone H3 lysine 27 (H3K27ac) coincident with the activation of targeted human promoters. In addition, we uncover new mediators of resistance to the BRAF V600E inhibitor PLX-4720 in human melanoma cells using genome-scale screening with dCas9-dMSK1. Collectively, our findings enable a new way to reshape human chromatin using CRISPR/Cas9-based epigenome editing and establish the causal link between histone phosphorylation and human gene activation.

Rational Design and Assessment of Protein-Lipid Hybrid Vehicle for Non-Viral Gene Delivery

Conference: International Conference on Epigenetics and Bioengineering

Year: 2020

Proceeding: 4th International Conference on Epigenetics and Bioengineering (EpiBio 2020)

Group: Poster Session

Session: Poster Session

Authors:

Thomas, J., NYU Tandon
Monkovic, J., NYU Tandon
Katyal, P., New York University Tandon School of Engineering
Punia, K., NYU Tandon
Montclare, J. K., New York University

Gene therapy has the potential to treat various diseases, but a vehicle capable of safely delivering nucleic acids does not currently exist for widespread clinical use. We have recently developed a lipoproteoplex (LPP) consisting of a super-charged coiled-coil protein (CSP) and a cationic liposomal carrier, that has the ability to condense nucleic acids and deliver them in vivo. A CSP library has been rationally designed to improve the efficacy of the LPP compared to the parent protein via increased nucleic acid binding and endosomal escape. The secondary structure and nucleic acid binding ability of each library member was assessed using circular dichroism and electrophoretic mobility shift assay, then compared to functional transfection data. Structural and functional data suggests that both high alpha-helicity of the protein component of the LPP as well as high nucleic acid binding affinity are crucial for efficient gene delivery into cells.

Regulatory Element and Cell-Type Specific Landscapes Control dCas9-Based Transcription Factor Efficacy in Human Cells

Conference: International Conference on Epigenetics and Bioengineering

Year: 2020

Proceeding: 4th International Conference on Epigenetics and Bioengineering (EpiBio 2020)

Group: Poster Session

Session: Poster Session

Authors:

Wang, K., Rice University
Li, J., Rice University
Hilton, I., Rice University

dCas9-based technologies have emerged as powerful tools to site-specifically write and erase endogenous epigenetic modifications and to modulate gene expression. However, a comprehensive comparison and benchmarking of dCas9-based transcriptional activators at cell-type specific human enhancers in different cell types is lacking. The functional evaluation of dCas9-based activators at enhancers is needed because enhancer function is tightly linked to cognate promoter activity (i.e. gene expression levels), developmental disorders, and human diseases. Here, we compare the efficacy of five state-of-the-art dCas9-based activators (dCas9-VPR, dCas9-SunTag, dCas9-SAM, dCas9-p300, and a novel dCas9-CBP fusion protein) at three testbed enhancer/promoter pairs across three human cell lines to evaluate the cell-type specific influence over their functions. Our results demonstrate that different dCas9-based activators have altered protein stabilities that vary based upon cell type, and furthermore, that the functional efficacy at enhancers and promoters is both locus and cell type specific. Furthermore, we conducted focused comparisons of theses activators spanning 1.5 kb over the OCT4 testbed promoter locus to illuminate how basal chromatin architecture impacts dCas9-based transcriptional activation, which reveals that underlying chromatin context impacts efficacy, in addition to cell-type specific milieu. Our comprehensive comparison of dCas9-based activators confirms the robust efficacy of our dCas9-CBP fusion protein at human enhancers streamlines the applicability of existing dCas9-based activators. Our work also provides insights that pave the way for future studies to define how cell-type specific *cis* (e.g. epigenetic context) and *trans* factors (e.g. protein complexes) influence both natural and synthetic transcriptional regulation.

Supporting Native Hawaiian Health with Community-Based Social and Biomedical Research

Conference: International Conference on Epigenetics and Bioengineering

Year: 2020

Proceeding: 4th International Conference on Epigenetics and Bioengineering (EpiBio 2020)

Group: Poster Session

Session: Poster Session

Authors:

Kunihiro, B., University of Hawaii at Manoa
Peres-David, R., University Of Hawaii at Manoa
Maunakea, A., University of Hawaii at Manoa
Wells, R., University of Hawaii at Manoa
McCracken, T., John A. Burns School of Medicine
Umeda, L., University of Hawaii at Manoa
Sultana, R., University of Hawaii at Manoa

PURPOSE: Research indicates that Native Hawaiians have the highest rates of diabetes and obesity in the state of HawaiÊ»i. Understanding the way epigenetics plays a role in Type 2 Diabetes (T2DM) will help us determine how environmental factors play a role in the risk of T2DM in Native Hawaiian populations. Our objectives with this study are:(1) Understand the relationship between social context and health; (2) Learn about the mechanism through which a community-based program impacted health/wellbeing of Native Hawaiians; (3) Consider a new model of community-engaged research that empowers individuals and organizations to address health disparities.

METHODS: The study targeted individuals (16 years or older) in the state of HawaiÊ»i affiliated with MAÊ»O organic farms or connected to their social network. Stool sample kits were distributed to participants with oral and written instructions to collect their own samples at home and store them in a freezer until they can be returned via mail or collected by a community research facilitator for analysis. Stool samples were used to metagenomics analysis. Height, weight, blood pressure, pulse and A1C were measured and recorded.

RESULTS: Results of this study show that main bacteria were dominant in Native Hawaiians (Bacteroidetes, Firmicutes, Proteobacteria, Actinobacteria). Actinobacteria and Proteobacteria were significantly correlated with BMI and A1c in Native Hawaiians. In the overall study population, there is a statistically significant negative correlation between A1c and the abundance of Actinobacteria (Ï□= -0.169; CI = 90%) and Proteobacteria (Ï□= -0.212; CI=95%). In the same group of participants, BMI is negatively correlated with the abundance of Actinobacteria (Ï□= -0.222; CI=95%). The results also show the impact of social networks on the health/wellbeing of the individuals.

Temporal Signaling and Information Processing in Chromatin-Mediated Gene Regulation

Conference: International Conference on Epigenetics and Bioengineering

Year: 2020

Proceeding: 4th International Conference on Epigenetics and Bioengineering (EpiBio 2020)

Group: Poster Session

Session: Poster Session

Authors:

Mukund, A., Stanford University
Bintu, L., Stanford University

Chromatin-mediated gene regulation is a key pathway cells use to regulate gene expression in response to internal and external stimuli; however, the full range of dynamic gene expression patterns achievable using chromatin regulation is unknown. Chromatin regulators (CRs) activate or silence genes in an all-or-none fashion and have distinct degrees of epigenetic memory; this behavior can be described via phenomenological models in which cells transition between active and silent states. Using this approach, we demonstrate that repressive CRs without long-term epigenetic memory can filter out high frequency noise, and can be used in autoregulatory loops to control population gene expression profiles. Repressive CRs with long-term memory encode the total duration of their recruitment in the fraction of cells irreversibly silenced, but cannot sustain oscillatory dynamics. Activating CRs can be used in combination with repressive CRs to produce well-known behaviors including adaptation and ultrasensitivity. Additionally, we quantify the ability of CRs to encode the duration of their recruitment in the fraction of cells silenced using an information-theoretic approach, and analyze the effect of input and output noise on the fidelity of CR signaling. Altogether, we describe dynamics of gene regulation achievable by CRs based on phenomenological models of chromatin regulation, demonstrate how they can be used to build complex signaling behaviors, and provide a mathematical formalism for analyzing their ability to transmit information in cells.

Dynamic Nucleosome Reorganization during the Pioneer Factor GATA3-Mediated Cellular Reprogramming in Breast Cancer

Conference: International Conference on Epigenetics and Bioengineering

Year: 2020

Proceeding: 4th International Conference on Epigenetics and Bioengineering (EpiBio 2020)

Group: General Submissions

Session: Sensing Epigenetic Modifications

Authors:

Takaku, M., University of North Dakota School of Medicine and Health Sciences
Saotome, M., University of North Dakota School of Medicine and Health Sciences

Transcription factors (TFs) recognize short DNA sequences to regulate gene expression. They play crucial roles in cell development and reprogramming. Eukaryotic cells have chromatin structure to package huge DNAs inside of the tiny nuclei. Therefore, TFs need to overcome this physical barrier to initiate cell differentiation and reprogramming. A subset of transcription factors, called pioneer factors, are thought to be special as they are capable of overcoming the chromatin structure by using their ability to bind to nucleosomes and induce chromatin opening. Although various pioneer factors have been identified in multiple tissues and human diseases, the fundamental mechanisms underlying their gene-regulatory actions remain elusive. Previously, we demonstrated that GATA3, one of the most frequently mutated genes in breast cancer, acts as a pioneer factor to suppress tumor growth in MDA-MB-231 human breast cancer cells by inducing mesenchymal-to-epithelial transition (MET). Importantly, although GATA3 is capable of inducing chromatin opening, we found that GATA3 can only open ~50% of closed chromatin binding sites suggesting that GATA3 action is chromatin context-dependent. Context-dependent actions of other pioneer factors have been also reported in the other experimental systems. Here, we developed a high-resolution nucleosome mapping technique (Capture MNase-seq) in human cells to capture dynamic nucleosome remodeling during the MET cellular reprogramming. Capture MNase-seq revealed unique nucleosome positioning at the pioneer factor binding sites, associated with gene activation activities of GATA3. We also found that the nucleosome formation on the pioneer binding sites is regulated by chromatin remodeling factors. The nucleosome mapping data enabled us to determine the Cryo-EM structure of the GATA3-nucleosome complex. This structure revealed that GATA3 uses its two zinc finger domains to stably bind to nucleosomes. These data suggest that the engagement of the pioneer factor in a specific spatial location on the nucleosome may dictate productivity of new enhancer formation.

Engineering Transcription Factors that Sense Histone Modifications in Cancer Cells

Conference: International Conference on Epigenetics and Bioengineering

Year: 2020

Proceeding: 4th International Conference on Epigenetics and Bioengineering (EpiBio 2020)

Group: General Submissions

Session: Sensing Epigenetic Modifications

Authors:

Haynes, K., Emory University

Mounting evidence from genome-wide comparisons of chromosome packaging and gene expression in healthy versus cancer cells suggests that epigenetic hyper-repression, rather than genetic mutation in many cases, supports cancer aggressiveness. Aberrant behavior of the chromatin system (genomic DNA, and nuclear RNA and proteins) has been implicated as a driver of metastasis and drug resistance. Since the early 1990â€™s small compounds have been used to disrupt hyper-repressed chromatin to simultaneously activate sets of therapeutic genes in cancer cells. However, it is difficult to customize the biological activity of these small compound inhibitors, and they do not directly mediate RNA PolII activity at silenced tumor suppressor genes. To address these limitations, our lab has designed synthetic reader-actuator (SRA) fusion proteins that bind epigenetic marks within chromatin. The first SRA we have designed and tested in cancer cells, â€œPolycomb-based transcription factorâ€□ (PcTF), reads histone modifications through a protein-protein interaction between its N-terminal Polycomb chromodomain (PCD) motif and trimethylated lysine 27 of histone H3 (H3K27me3). The C-terminal VP64 domain of PcTF recruits endogenous activators to silenced targets. We identified a set of 104 genes that become consistently activated from 24 - 72 hours after PcTF-overexpression in a triple negative breast cancer cell line (BT-549). PcTF-positive BT-549 cells show reduced proliferation and viability in vitro. We propose that SRA-responsive genes, which includes anti-proliferation, immune response, and epithelial cell fate genes, are part of a â€˜druggableâ€™ anti-cancer gene module. These genes are rarely mutated in basal tumors and therefore may produce a therapeutic effect when they become activated. This talk will review our published results for BT-549, new data from additional TNBC lines, progress in engineering enhanced SRA activity, and analysis of 99 genes that are repressed in both cell lines and basal tumors to support clinical translation of our research.

Coordinated Regulation of DDIT4 Transcriptional Heterogeneity By a Cluster of Glucocorticoid Receptor Binding Sites

Conference: International Conference on Epigenetics and Bioengineering

Year: 2020

Proceeding: 4th International Conference on Epigenetics and Bioengineering (EpiBio 2020)

Group: General Submissions

Session: Single Cell Epigenetics

Authors:

Hoffman, J. A., National Institute of Environmental Health Sciences
Day, C. R., National Institute of Environmental Health Sciences
Rodriguez, J., National Institutes of Environmental Health Sciences
Trotter, K. W., National Institute of Environmental Health Sciences
Archer, T. K., National Institute of Environmental Health Sciences

Synthetic glucocorticoids are widely prescribed drugs used to treat a variety of human diseases including auto-immune diseases, asthma, cancer, and COVID-19. The transcriptional response to glucocorticoids such as Dexamethasone (Dex) is elicited by the Glucocorticoid Receptor (GR). Upon Dex treatment, GR enters the nucleus and interacts with thousands of enhancer elements throughout the genome. We recently demonstrated that the Dex response in human breast cancer cells is highly heterogeneous and that individual cells have unique transcriptional responses to Dex. To examine whether this heterogeneity arises from differential utilization and activity of GR-bound enhancers, we focused on the Dex response at the GR target gene DNA Damage Inducible Transcript 4 (DDIT4). Using ChIP-seq, ATAC-seq, and RNA-seq we identify 4 potential enhancers with GR binding sites (GBSs) 18-30kb upstream of the DDIT4 TSS and differential patterns of chromatin accessibility, histone acetylation, SWI/SNF recruitment, and eRNA transcription. We generated homozygous deletions of each GBS with CRISPR/CAS9 and used single molecule FISH (smFISH), RT-PCR, and ChIP-seq to determine their unique requirements. While GR binding to these GBSs was independent, each GBS had unique effects on histone acetylation, DDIT4 transcription, and the activity of the other GBSs/enhancers. Three of the GBSs were required for the timing and/or degree of DDIT4 induction. Conversely, the fourth GBS acted as a suppressor and deletion resulted in significant up-regulation of both DDIT4 transcription and enhancer RNA transcription from the other GBSs. Strikingly, smFISH revealed that these enhancers contributed to cellular heterogeneity, as deleting GBSs altered the frequency and amplitude of DDIT4 transcription across cell populations. Taken together, these results demonstrate that individual GBSs uniquely contribute to cell-to-cell heterogeneity within the transcriptional response of DDIT4 to Dex. Furthermore, they underscore the possibility that targeted modification of individual GBSs could be utilized to tailor patient-specific strategies for treatment of human diseases.

Detangling DNA: Single-Cell Approaches to Understanding Plasticity in Cellular Reprogramming

Conference: International Conference on Epigenetics and Bioengineering

Year: 2020

Proceeding: 4th International Conference on Epigenetics and Bioengineering (EpiBio 2020)

Group: General Submissions

Session: Single Cell Epigenetics

Authors:

Galloway, K., Massachusetts Institute of Technology

Cellular reprogramming continues to generate new cell types, increasingly expanding our perspective of cellular plasticity. Despite improved genetic tools and epigenetic modulations, reprogramming remains a rare cellular event. Recent studies in reprogramming indicate that changes in DNA topology precede changes in gene expression, suggesting that transcriptional and epigenetic realignment require particular 3D genome conformations. In this talk, I will describe how I identified molecular roadblocks in reprogramming that arise from tradeoffs between transcription and proliferation rates. Utilizing single cells RNAseq (scRNAseq), I recently identified topoisomerases, enzymes that curate DNA topology, as regulators of cellular reprogramming. Supported by topoisomerase activity, elite cells maintain high rates of transcription and replication to reprogram at near-deterministic rates. My discovery highlights how topological stress impacts the function of gene networks (e.g. native or synthetic circuits) and constrains cellular transitions. I will discuss how this finding opens completely new questions about how the structure of the genome stabilizes cellular identity and suggests strategies to improve the design of synthetic gene circuits.

Environmental Regulation of the Estrogen Response in Single Cells

Conference: International Conference on Epigenetics and Bioengineering

Year: 2020

Proceeding: 4th International Conference on Epigenetics and Bioengineering (EpiBio 2020)

Group: General Submissions

Session: Single Cell Epigenetics

Authors:

Rodriguez, J., National Institutes of Environmental Health Sciences

Gene regulation is complex, involving the coordination of hundreds of proteins, including chromatin remodelers, enhancer RNA and chromosome looping. The estrogen response has served as a model of gene regulation due in part to its rapid and extensive induction. Single molecule RNA imaging in living cells has shown that this process is dynamic and heterogenous in time leading to differences in cell responsiveness and expression in the cell population. This variability in expression at the tissue and single cell level increases with age, implicating environmental epigenetic regulation. Environmental factors such as endocrine disruptor chemicals are found in our food and can disrupt the normal estrogen response. However, how single cells respond to endocrine disruptor chemicals is unclear. To investigate the interplay between the environment and the estrogen response we are using single-molecule imaging of the estrogen receptor and transcription in living cells. This work will reveal mechanistic insights into how environmental factors perturb estrogen receptor mediated transcription.

Tunable Optogenetic Signal Decoding through Epigenome Editors

Conference: International Conference on Epigenetics and Bioengineering

Year: 2020

Proceeding: 4th International Conference on Epigenetics and Bioengineering (EpiBio 2020)

Group: General Submissions

Session: Single Cell Epigenetics

Authors:

Lee, J. B., North Carolina State University
Caywood, L., NC State University
Keung, A. J., North Carolina State University
Levering, N., North Carolina State University
Lo, J. Y., North Carolina State University

In eukaryotic cells, access of transcription machinery to genomic binding sites is controlled by chromatin structure, which can be modified by various epigenome editors (EEs). EEs are used for various applications including high-throughput screens for cell phenotypes and drug targets, cell differentiation, and mechanistic studies of epigenetic regulation. EEs have mostly been used bimodally to activate or repress gene expression; however, by utilizing dynamic signaling, they could be used to execute a larger landscape of gene outputs in terms of expression level and information transmission. Inspired by natural systems that use transcription factor dynamics to encode different gene expression programs, we developed an optogenetic system to dynamically recruit a representative EE, VP16, to a fluorescent reporter in *Saccharomyces cerevisiae*. The input signal was modulated via frequency, intensity, and pulse width of light pulses. This system resulted in tunable expression levels that spanned two orders of magnitude. Additionally, we quantified the reliability of this system at transmitting information using the maximal mutual information (MI). We found that frequency modulation was the most reliable mode of modulation; however, it was limited to three error-free states (1.6 bits). To determine how chromatin state affects MI and expression levels, we modulated the chromatin state by constitutively recruiting a library of over 100 orthogonal EEs. A handful of the EEs increased the MI, and many decreased the MI. Furthermore, we found some of the EEs are able to alter the input-output transfer function. For example, many of the EEs exhibited band-pass filtering. We conclude that dynamic recruitment of EEs can increase the tunability of expression levels but are limited in the amount of information that can be reliably transmitted. However, altering the chromatin state can increase the information capacity and also provide complex signal processing behavior.

Cooperativity and the Design of Gene Regulatory Circuits

Conference: International Conference on Epigenetics and Bioengineering

Year: 2020

Proceeding: 4th International Conference on Epigenetics and Bioengineering (EpiBio 2020)

Group: General Submissions

Session: Writing Epigenetic Modifications

Authors:

Khalil, A. S., Harvard University

The establishment and heritable propagation of gene expression states are defining features of multicellular development and are central to adaptation to new environments. Studies across eukaryotes have uncovered key molecular regulators and mechanisms involved in these processes, and produced a picture of transcription and epigenetic regulatory â€œcircuitsâ€☐ with remarkable properties. In this talk, I will describe how we are using synthetic rational design approaches to illuminate design principles of these systems. Using this approach, we design and construct regulatory circuits in eukaryotic cells from the bottom-up. This allows us test different circuit designs, with precisely tailored biochemical parameters, to identify which circuits are sufficient to program regulatory function. We have been particularly focused on applying this approach to recapitulate salient features of natural chromatin-based epigenetic systems and accordingly to decipher basic requirements of epigenetic memory. We also hope that this approach will help to establish a powerful foundation for the purposeful manipulation of epigenetic regulatory systems to combat disease and program new traits in cells and organisms.

Epigenetic Editing Regulates Alternative Splicing of Srsf11 and Cocaine Reward Behavior

Conference: International Conference on Epigenetics and Bioengineering

Year: 2020

Proceeding: 4th International Conference on Epigenetics and Bioengineering (EpiBio 2020)

Group: General Submissions

Session: Writing Epigenetic Modifications

Authors:

Heller, E. A., University of Pennsylvania
Xu, S. J., University of Pennsylvania
Lombroso, S.

Cocaine drives alternative splicing to a far greater extent than differential mRNA expression yet this mechanism of gene regulation is understudied in the context of drug abuse and addiction. During development, alternative isoform expression confers neuronal identity and is maintained throughout life. We posit that in adulthood, alternative exon expression maintained by stably-associated hPTMs, underlies chronic disease states, such as addiction. We and others have recently reported that alternative splicing is correlated with H3K36me3 enrichment in brain, a histone posttranslational modification (hPTM) which is also regulated by cocaine. H3K36me3 is catalyzed by the histone methyltransferase, SET2. We generated three within-sample, RNA/ChIP datasets from NAc of mice following either SET2 overexpression, 1-day or 28-days of abstinence following cocaine-self administration. In each dataset, we analyzed alternative isoform expression and splice junction enrichment of H3K36me3. To examine the direct causal relevance of H3K36me3 to alternative splicing, we expressed either full-length SET2 or locus-targeted dCas9-SET2 in NAc (control: catalytically dead SET2(R195)). Across three datasets, we identified exons that were both differentially expressed and enriched for H3K36me3 at the respective splice junction (FET $P<0.05$). This group included Serine and arginine rich splicing factor 11 (Srsf11). We selected this gene given that the Srsf11 target motif was enriched globally in spliced genes following cocaine- and SET2-treatment. Targeted enrichment of H3K36me3 by dCas9-SET2 to the relevant *Srsf11* splice junction was confirmed by CUT&RUN-seq in N2a cells. This manipulation increased inclusion of the targeted skipped exon and increased cocaine conditioned place preference and self-administration behavior. Taken together, the combination of bioinformatic and epigenetic editing approaches provide conclusive evidence of the functional relevance of H3K36me3 enrichment at splice junctions to expression of specific isoforms.

High-Throughput Discovery and Characterization of Human Transcriptional Effectors

Conference: International Conference on Epigenetics and Bioengineering

Year: 2020

Proceeding: 4th International Conference on Epigenetics and Bioengineering (EpiBio 2020)

Group: General Submissions

Session: Writing Epigenetic Modifications

Authors:

Tycko, J., Stanford
Bintu, L., Stanford University

Josh Tycko, Nicole DelRosso, Gaelen T. Hess, Aradhana, Abhimanyu Banerjee, Aditya Mukund, Mike V. Van, Braeden K. Ego, David Yao, Kaitlyn Spees, Peter Suzuki, Georgi K. Marinov, Anshul Kundaje, Michael C. Bassik*, Lacramioara Bintu*

Thousands of proteins localize to the nucleus, however it remains unclear which contain transcriptional effectors. Here, we develop HT-recruit - a pooled assay where protein libraries are recruited to a reporter, and their transcriptional effects are measured by sequencing. Using this approach, we measure gene silencing and activation for thousands of domains. We find a relationship between repressor function and evolutionary age for the KRAB domains, discover Homeodomain repressor strength is collinear with Hox genetic organization, and identify activities for several Domains of Unknown Function. Deep mutational scanning of the CRISPRi KRAB maps the co-repressor binding surface and identifies substitutions that improve stability/silencing. By tiling 238 proteins, we find new repressors as short as 10 amino acids. Finally, we report new activator domains, including a divergent KRAB. Together, these results provide a resource of 600 human proteins containing effectors, and demonstrate a scalable strategy for assigning functions to protein domains.

Zfp462 Recruits HP1 and Ehmt1/2 to Repress Endodermal Genes in Neuronal Cells

Conference: International Conference on Epigenetics and Bioengineering

Year: 2020

Proceeding: 4th International Conference on Epigenetics and Bioengineering (EpiBio 2020)

Group: General Submissions

Session: Writing Epigenetic Modifications

Authors:

Bell, O., Keck School of Medicine of the University of Southern California

Haploinsufficiency of ZNF462 has recently been shown to cause Weiss-Kruszka Syndrome, a complex neuro-developmental disorder characterized by a spectrum of neurodevelopmental defects including Autism. ZNF462 is a conserved, putative transcription factor that is essential for embryonic development. However, its precise roles in transcriptional regulation and development are not understood. Using a CRISPR genetic screen for regulators of Heterochromatin Protein 1 (HP1)-dependent silencing in mouse embryonic stem cells (mESCs), we discovered that murine homolog Zfp462 is involved in heritable transcriptional silencing through protein interactions with H3K9-specific HMTase G9a/GLP and with Heterochromatin Protein 1 (HP1) isoforms. Zfp462 occupies genomic loci associated with transposable elements (TEs) harboring DNA binding sites for ESC- and endoderm-specific transcription factors (TFs) that are critical for tissue-specific gene expression. Zfp462 deletion in mESCs results in increased chromatin accessibility at target sites and aberrant expression of endodermal genes during neuronal differentiation. We show that Zfp462 controls TE-derived endoderm enhancers by targeting heterochromatin modifiers to restrict TF binding and activation. Together, our data suggest that aberrant activation of endodermal genes interferes with neuronal lineage-specification which may underlie the neurodevelopmental pathology.

AUTHOR INDEX